FANTAIL BOOKS

Free Willy:
The World of Killer
Whales

Nigel Robinson
Free Willy:
The World of Killer Whales

Based on the film: Free Willy
Screenplay by Keith A. Walker *and*
Corey Blechman
Story by Keith A. Walker

FANTAIL

FANTAIL BOOKS

Published by the Penguin Group
Penguin Books Ltd, 27 Wrights Lane, London W85TZ, England
Penguin Books USA Inc., 375 Hudson Street, New York, New York, 10014, USA
Penguin Books Australia Ltd, Ringwood, Victoria, Australia
Penguin Books Canada Ltd, 10 Alcorn Avenue, Toronto, Ontario, Canada M4V 3B2
Penguin Books (NZ) Ltd, 182–190 Wairau Road, Auckland 10, New Zealand

Penguin Books Ltd, Registered Offices: Harmondsworth, Middlesex, England

Published in Fantail Books 1994
3 5 7 9 10 8 6 4

FREE WILLY, characters, names, and all related indicia are trademarks of Warner
Bros. © 1993 Warner Bros., Regency Enterprises, Studio Canal + .
All rights reserved
Fantail Film and TV Tie-in edition first published 1994

Typeset by Datix International Limited, Bungay, Suffolk
Filmset in 14/16 pt Monophoto Times
Printed in England by Clays Ltd, St Ives plc

Contents

Introduction

The peaceful ocean – any ocean – on a hot summer's day. There is no sign of land; in fact, there's nothing to be seen for miles around, only the shimmering blue line of the far-away horizon. The sun is glittering down on the water which is gently lapping against the side of your boat.

It's perfectly silent. There is no sound apart from the soft whispering of the waves and the creaking timbers of your tiny boat. You know that, deep down below you, there are countless hundreds of millions of fish. But up here, you're on your own; alone with nothing for company but your boat,

1

the sun and the cloudless sky, and the wide, fathomless and formless ocean.

And yet still you have the curious feeling that you're being watched. You've felt that way ever since you left harbour this morning. As if someone was following you, watching your every move.

Crash!

You turn around to the starboard side of the boat. The sound is deafening, like the sound of someone diving into the sea only louder. What is it? Has the boat hit something? But there's nothing, only the white foam on the ocean waves. Maybe it was your imagination but you could have sworn . . .

Crash! There it is again, but this time to port. You rush over to the portside of the boat, and then you see them.

There must be five of them. No, ten. Twelve. Maybe even as many as twenty. There are so many you lose count. The air explodes with the sound of the creatures leaping out of the water. Their long powerful bodies glisten with muscle, as they execute a perfect arc, before crashing back down under the waves again. All of them in perfect formation, as if they were taking part in a well-choreographed ballet.

Killer whales! The devils of the deep! Your heart sinks. You've heard all the stories about them being the most ferocious beasts in the entire ocean. Why else would they be called killers? They must have scented your blood and been following you since early morning, when you set sail on this short journey down the coast.

Killers! You're done for. It's only a matter of time now before they capsize your tiny boat and drag you under with them, with those vicious white teeth. A human being would make a tasty afternoon snack for them.

And yet . . . If they've really been following you since early this morning, why haven't they attacked you before now?

Then the strangest thing happens. Almost as one, the whales rise out of the water and for a few seconds hold themselves up vertically, treading the water with their mighty tails. They're all looking at you with those tiny eyes with the distinctive white markings.

Suddenly you're no longer afraid, as you

recognize that look in their eyes. They're curious, that's all – they just want to see who you are. They bare their teeth but they no longer look vicious and threatening; if you didn't know better, you'd say that they might even be grinning.

For a further few seconds you continue to look at each other. You recognize a glimmer of something else in their eyes too . . . Could it be that they're sizing you up, deciding whether or not you're a threat to them? Could it be that they're actually *thinking*? Of course not. They're only fish, *aren't they?*

Then suddenly, as if on a prearranged signal, the killers dive back into the ocean again and swim swiftly away, disappearing into the distance. You breathe a sigh of relief and then wonder what there is to be so relieved about? After all, did they attack you? Did they try to capsize your boat?

It's not as if you've never seen a killer whale before, is it? Didn't your parents take you to that marine theme park when you were a young kid? You saw a killer whale then, didn't you – swimming alone in its tank, as hundreds of tourists and sightseers waited for it to perform its spectacular tricks . . .

Yet even as a young kid you felt that there was something sad and mournful about that killer whale in its tank. And it's not until now, now that you've seen the sheer exhuberance and the freedom enjoyed by that group of whales, that you realize what the problem was. He was homesick, homesick for the wild, open oceans and the company of his own brothers and sisters . . . his family – just like the killer whale in the film *Free Willy*.

1

Out of the Water – and Back
Again

The earth on which we live isn't the earth at
all. Whoever gave it its name should really
have called it the ocean, for over 70 per
cent of our planet's surface is covered with
water.

From the smallest off-shore bays to the
mighty Pacific, the largest ocean in the
world; from the tropical waters of the South
Seas to the frozen ice floes and icebergs of
the Arctic wastes, almost three-quarters of
the earth's surface is covered with water.

Man has conquered much of the land.
But there are still places which are as un-
known to him as the farthest stars and
galaxies of outer space. Neil Armstrong

landed on the moon almost twenty-five years ago in 1969, but today scientists probably know more about our nearest neighbour in space than they do about the mysteries of the oceans.

For something which is literally vital to our existence – life on earth began many billions of years ago in the sea, after all – the oceans of the world are surprisingly inhospitable to man. If a diver with just a diving suit and an oxygen supply were foolhardy enough to venture down unprotected to the deepest parts of the ocean he'd be crushed to death in seconds.

That is, of course, if he didn't freeze to death first. While ships regularly break through the icy waters of the Arctic, where temperatures can drop to as low as minus 80 degrees Celsius (that's about minus 120 degrees Fahrenheit), a man left alone in the sub-zero waters of the Arctic in the north, or the Antarctic in the south, would stand little chance of surviving for even a few minutes.

Yet there's not a single place in the ocean that isn't teeming with life of one kind or another. Every species living there survives, with apparent ease, in conditions which would kill a mere human being.

In the deepest part of the Pacific Ocean, so deep that the light from the sun never reaches it, strange transparent fish, like something from a science-fiction film, flit about in the coldest depths, their own phosphorescence providing the light which enables them to see in the pitch blackness.

But of all the creatures living in the oceans of the world, there are probably none as graceful or as intelligent as the whales and the dolphins.

Darting through the waters of the world, from the frozen Arctic to the tropical Caribbean, they can reach incredible speeds. The killer whale, for instance, can swim at a maximum speed of twenty-seven knots per hour – that's about fifty kilometres per hour – and can easily outrace any fish.

However, just as the earth shouldn't really be called the earth, so whales and dolphins aren't fish! Despite their appearance, they are mammals just like ourselves. If you wanted to stretch a point, you could say that men and whales are very distantly related to each other!

*

About 3500 million years ago all life as we know it began in the ocean. Microscopic bacteria and algae appeared, feeding on the rich nutrients and chemicals in the waters of the newly cooled planet. From these simple organisms more complex creatures developed: plankton, simple worms and finally fish. For much of the earth's history the only form of life was to be found in the sea. As life developed, so the number of fish swimming in the sea increased.

With these growing numbers came a correspondingly growing demand for precious life-giving oxygen. Some fish realized that there was an abundant source of oxygen, and food, which had not yet been exploited; the dry land was waiting to be colonized.

Over millions of years, some fish managed to adapt themselves to life out of the water. Over thousands of generations they lost their fins, which changed into legs. Constantly adapting to changing circumstances, they became the first amphibians, capable of living both in the water and on dry land. Turtles are a good example of modern-day amphibians.

Thousands upon thousands of years

9

passed and many more species developed from those first adventurous creatures who had dared to leave the water. The dinosaurs appeared, as well as the early mammals – tiny shrew-like creatures – our ancestors. The dinosaurs died out about sixty-five million years ago (no one knows exactly why; some scientists think it's possible the earth collided with a giant meteor, others that there was a dramatic change in the earth's climate), and afterwards mammals were free to become the dominant species on the earth. As time passed, so one mammal grew to be the ruler of the planet; that mammal was man.

It's easy to see the similarities between a man and, for example, a chimpanzee or a gorilla. We've all got four limbs, are roughly the same shape, and have distinct facial resemblances. But what do humans and whales have in common? At first sight they couldn't be more different. Let's look more closely.

The most important distinction between mammals and other species is the way in which they give birth to their young. Fish, birds and reptiles all lay eggs which are then incubated until the young hatch out of them. Mammals, on the other hand, give birth to live young. Human mothers do this. So do chimpanzees, lions and dogs, for example. This is also exactly what whales and dolphins do. Like humans, chimps, lions and dogs, whales and dolphins suckle their young, feeding them, at least for a time, on warm milk.

Fish breathe through their gills, the narrow slits at the sides of their heads. Whales, however, breathe through their lungs, just as you or I do. This means they must come up to the surface every so often to breathe. In fact, the most common cause of death amongst whales and dolphins – once you've excluded being killed by man – is drowning.

Because air is so important to them, the 'nostrils' of a whale are not quite where you'd expect them to be. Instead of being in the front of their 'face' between their eyes and above their mouth as ours are, they're to be found on the top of their head – the side nearest to the surface of the ocean, and so the side which breaks through the water first when the whale surfaces from the depths.

In many whales, the two nostrils have been replaced by one single blowhole or spout. When the whale reaches the surface it breathes out an enormous amount of used air – wouldn't you if you had been holding your breath for so long? When the expelled warm breath comes into contact with the cooler surrounding air it turns into vapour. To any passing sailor this would look just like a spout of water. That's the origin of the popular myth that whales and dolphins expel water through their blow-holes. It's estimated that every time a whale surfaces it exhales about 90 per cent of used air, compared with the 15 per cent a human swimmer exhales.

There are other similarities between whales and humans. Some of the larger whales even have traces of whiskers around their heads and snouts, the only hairs on an otherwise completely smooth and hairless body. It's thought that these whiskers are particularly sensitive and act in much the same as a cat's whiskers.

If you're in any further doubt that whales and dolphins are mammals and not fish, you have only to look at a picture of a growing whale embryo. Shortly after it's been conceived and is still growing in its mother's womb it looks remarkably like a human foetus at the same relative age. You can pick out what will later be its head, tucked tightly into itself. You will also see what look very much like arms and legs. These eventually become fins, but it's conclusive proof of a whale's relationship with other mammals and the fact that, long ago, whales and men shared a common ancestor.

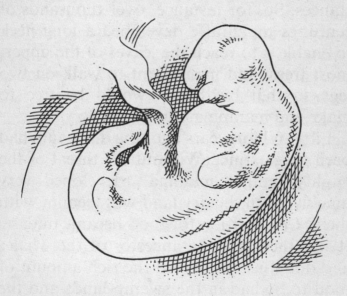

So whales and dolphins are definitely mammals and not fish. Perhaps that might explain why humans and whales can often get on so well together. But there are many other mammals which live in or near the sea – sea-lions, seals and penguins, to name just a few – and none of these could ever be mistaken for fish. Why do whales and dolphins look so much like the fish they're often taken to be?

The answer is simple and lies in the workings of evolution. To survive, each animal must adapt itself to its individual circumstances. So, for instance, over thousands of centuries the giraffe developed a long neck to enable it to reach the leaves of the uppermost trees; and man learnt to walk on two legs so that his hands would be free to make and manipulate tools.

Fifty million years ago, the dinosaurs had been long extinct. Without the threat of the mighty reptiles mammals developed very quickly, until the dry land was teeming with them. Competition for food became intense. It was then that the ancestor of the whales and dolphins discovered the rich amount of food to be had in the swamp-lands and the marshes.

As millions of years passed, so its body

14

changed, enabling it to take more advantage of this valuable new food source. Its nostrils moved towards the top of its head, enabling it to breathe when its head was lowered underneath the water. It developed flippers to help it move about in the water much more easily, and began to take advantage of the rich food to be found in the shallow waters on the coast where few other mammals went. Like today's otters, for a long time it divided its time equally between the land and the water.

Then finally this mammal made the decision to spend its entire life in the ocean. As the years passed, so its body changed even more, adapting itself to the water. It lost most, and in some cases all, of its body hair: hair slows an animal down in the water (that's why champion swimmers shave their whole bodies), and this creature had to be speedy to hunt the fast-moving fish in the sea. It developed a tail because a strong tail is much better for propelling you through the water than legs; its 'forearms' became paddle-shaped flippers which it used to steer and stabilize itself in the sea; and it developed a streamlined body,

thereby reducing even further the water resistance.

The reason why whales and dolphins look like fish is very simple: the 'body' of a fish is by far the best shape for living in the sea. And of all the mammals who live or swim in the sea – such as seals, polar bears and even man himself – it's the whales and dolphins who have had the greatest success in adapting themselves to their new environment. Like no other mammals before them, they have reconquered the sea and, in doing so, have established themselves as kings of the waves. But as you can see in the film, that hasn't stopped man intruding into the killer whales' kingdom.

2

When is a Whale not a Whale? When it's a Killer

Killer whales, like Willy, belong to a large family of marine mammals known as cetaceans. This family includes, amongst other, the mighty blue whale, the largest animal in the world, measuring some 110 feet, right down to the dolphins and the porpoises you might see at an aquarium or zoo.

While the killer whale isn't anywhere near the size of the blue whale, it's still the largest of the dolphin-like group of cetaceans and can grow to a size of almost ten metres (about thirty-two feet). The usual size, however, is about eight metres for a male, and about seven metres for a female. Baby killer whales, or calves, are usually about two

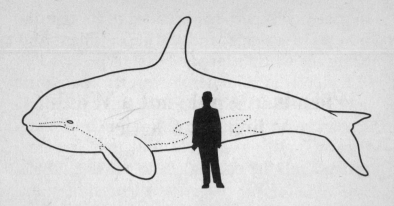

metres at birth and when they are mature can weigh up to 10,000 kilograms (that's about ten tons)!

Just as it is incorrect to call a whale a fish, it is also wrong to describe animals like Willy as whales. It's best to think of killer whales as being a type of dolphin, although they differ from the common dolphin in several respects. For instance, the head of a killer whale is blunt and isn't beaked like the common dolphin's.

Strictly speaking, though, the killer whale belongs to a separate species called orcinus which was first identified in 1860. The killer whale's scientific name is *Orcinus orca*, although most people today refer to them just as orcas. The Romans, however, knew it as the sea devil, perhaps because of its reputation as a savage and ruthless killer, and being the world's largest beast of prey.

Killer whales generally prefer to live in coastal areas such as Patagonia in South America, or off British Columbia and Vancouver Island in Canada. But they live and thrive in all the oceans of the world, and it's entirely possible that you might spot some off the coast of Britain if you're lucky. They're particularly fond of the North Sea and the Atlantic Ocean.

Just as South Americans are different from Canadians, so orcas from one part of the world differ slightly from those from another area. The orcas who swim in the cold waters of the Antarctic, for example, are smaller than others and have different feeding habits and behaviour patterns. It's even been suggested that these Antarctic whales should form an entirely separate species. However, at the moment scientists formally recognize only one race of killer whales whether they live in the waters around British Columbia or the coastal waters near Japan.

There's nothing to equal the thrill of

seeing a whole group of orcas, often as many as twenty or thirty, charging and splashing through the waves in perfect synchronization, and leaping clear out of the water with all the grace and seeming ease of a top-class ballet dancer. It's an unforgettable sight, and something which you would never see in an oceanarium or zoo where the animals are confined to relatively small water tanks and rarely, if ever, enjoy the freedom of the open sea.

So, supposing you were lucky enough to come across a group of killer whales – or orcas – in their natural habitat, how would you recognize them?

An orca is one of the simplest dolphins to identify, largely because of the big fin on its back. This dorsal fin, which can be as tall as two metres in a male, serves as a stabilizer. Often only this fin is visible above the surface of the water, and you'd be forgiven for thinking that what you'd spotted in the water was a shark.

The fin is shaped like a sword, which is probably why over the years the orca has been known by other names such as the sword of the sea, the gladiator whale and the swordfish (although the real swordfish is a different animal altogether, and a genuine fish). This fin looks so sharp and vicious that many people believe that the whale uses it to slash open the bellies of its victims, although there's no scientific evidence for this. The dorsal fin of a male orca usually stands firm and erect, while a female's may curve slightly.

Like Willy, orcas are smooth all over and their basic colour is black. They also have distinctive white markings on their bellies and flanks, and above and behind each of their eyes. These eye-patches can often resemble horns which might explain the Romans' sinister name for orcas. No two killer whales are alike and researchers use these markings, as well as nicks and cuts in their dorsal fins, to distinguish individual whales, perhaps even years after first spotting them.

There is another species – the false killer whale – which looks very similar to the orca. This imposter, however, is black all over with no white markings at all. It's highly unlikely that you'll ever see one in its natural habitat as, unlike most orcas, they prefer living out in the middle of the oceans rather than nearer the coast.

Killer Whale Fact File

• *Amongst the larger whales there are two distinct species: the toothed whales, such as the blue whale, dolphins and killer whales like Willy; and the baleen, or whalebone, whales, which have no teeth. The toothed whales feed mainly on fish and, in the case of the killer, anything which moves in the sea! The baleen whales, unable to chew, live mainly off krill, small shrimp-like crustaceans, which they sift through the horny strips of baleen lining their mouths.*

• *One of the earliest reports of killer whales in Great Britain was in 1691, when an entire pod was stranded on the shore of the Firth of Forth.*

• *The longest-lived killer whale, if we can believe the records, was 'Old Tom'. He was seen every winter in New South Wales, Australia, from 1843 to 1930, a total of eighty-seven years – longer than the average lifespan of a human being at that time.*

• *Killer whales usually prefer swimming in shallow waters, but one has been reported as diving to a depth of 300 metres.*

3

Riding with the Killers

The Tinglit Indians say that the first killer whale was born thousands of years ago when Natsilane, an Indian brave, who had become separated from his fellow braves, carved the first orca out of a lump of wood. The orca came to life and Natsilane rode on the mammal's back all the way home.

The American Indians have always held the orca in great reverence, and there exists even to this day a killer whale cult. But the orca has featured in the legends and folk-tales of many other countries as well.

In Norway witches were said to ride through the waves on the backs of orcas. Strangely enough, the killer whales were also regarded as lucky creatures, and if anyone killed an orca they could expect all manner of misfortunes to happen to them. (Sadly this hasn't stopped the mass slaughter of the orca's cousins, the larger whales.) If a group of whales, however, appeared in a place where they were seldom seen, that was considered a bad omen.

The Slavic peoples in Eastern Europe even believed that the world rested on the back of a whale, and earthquakes were said to be the consequence of this 'world-whale' stirring in its sleep. (Other people believed that the world rested on the back of an elephant, or even a tortoise, so perhaps we shouldn't take their claim too seriously!)

The legends of ancient Greece are also full of stories of men who have befriended orcas. The creature and its dolphin cousins were considered to be sacred to the sea god Poseidon. The Greeks believed that dolphins were either gods in disguise or the reincarnation of human souls, so to kill one was to invoke the wrath of the gods. The Greeks certainly recognized their great intelligence and there are many stories of their co-operation with human beings.

One famous tale tells the story of Arion, who actually lived in the seventh century BC. He was an accomplished musician who played at the court of Periander, the King of Corinth, in southern Greece.

Arion travelled to a prestigious music festival taking place on the island of Sicily, where he enchanted everybody present with the music he played on his lyre. The judges decided that he was by far the best musician at the festival and, in honour of his talent, he was presented with a small fortune in precious jewels and metals. Arion hired a boat and crew to carry him back to Corinth in triumph, and left the island a very rich man indeed.

Once they were safely out at sea, the crew mutinied and decided to murder Arion and steal his newly won jewels and gold. As a last request the musician asked that he be allowed to play one final song on his lyre. The cut-throats agreed and Arion changed into his finest robes and performed for the sailors on the open deck. His music so enchanted them that they almost spared his life, but the lure of treasure was too strong and they threw the musician overboard.

Arion's music had also attracted a school of orcas (like seals, whales have always been rumoured to have a fine appreciation of music!), and one of them rescued the drowning musician and allowed him to ride on his back all the way home to Corinth.

When the murderous sailors returned home the king asked what had befallen Arion. The sailors claimed that Arion had decided to remain behind in Sicily, at which point the King of Corinth saw through their lie and ordered their execution.

The orca which saved Arion remained in Corinth but, perhaps like its modern-day counterparts, it didn't take to captivity too well, and died shortly afterwards. Arion was so grief-stricken at the death of the orca which had saved his life that he arranged a splendid funeral for it in Corinth!

Another band of sailors captured a wandering beggar and decided to sell him into slavery. What they didn't know, however, was that their prisoner was the god Dionysius. When they discovered his true identity they were so overcome with fear that they dived overboard, upon which the god turned them into the first dolphins and handed them over to the sea god to draw his chariot through the waters for ever.

People say that a dolphin was also responsible for saving the life of Icadius, a sailor shipwrecked in Italy. According to the story, the dolphin (who was really Apollo, the god of prophecy, in disguise) guided Icadius to Delphi in Greece – the name of the town comes from the Greek word for dolphin, *delphinus*. There is also another story in which a dolphin, or maybe an orca, rescues the Greek maiden Phineis who had been thrown into the sea as a sacrifice to one of the sea goddesses.

Pliny, the Roman writer who lived almost

two thousand years ago, tells of the relationship between a young boy and an orca, very similar to the friendship which developed between Jesse and Willy. The two became such great buddies that the orca would carry the young boy to school each day across the Bay of Naples, and be waiting for him later in the day for the return journey!

Whales of a larger sort, like the great blue whale, have also been responsible for many legends. The Bible tells the tale of the prophet Jonah who was thrown overboard from a ship and swallowed by a whale, in whose belly he survived until he sprang out of its mouth and up on to dry land. People also say that Alexander the Great descended to the bottom of the sea in a diving bell, where he encountered a whale so huge that it took three days to go past him; but as the longest whale ever recorded was only thirty-three metres long, perhaps the historians of the time exaggerated a little!

There are also numerous stories of sailors landing on small, undiscovered islands, on which they pitch camp and light a fire. It's only when it's too late that the 'island' turns out to be a giant sleeping whale which, awoken by the fire, turns over and drowns all the sailors!

Since they were first discovered in the Stone Age, and early man painted their likeness on the walls of caves, whales and dolphins like Willy have inspired man's imagination and proved to be one of his most reliable allies in the animal kingdom. As the relationship between Jesse and Willy shows, in many of those legends there's a lot of truth.

Killer Whale Fact File

• *The false killer whale, similar to the orca but with no white markings, was thought to be extinct until the mid-eighteenth century. The only evidence of its ever having existed was fossils, and scientists thought it had died out until a group of them were washed up on the shore.*

• *The largest killer whale ever held in captivity was Orky, who was on display to tourists in California. He measured eight metres in length and weighed 6350 kilograms!*

• Naturalists now think that there may be two separate races of killer whale. 'Resident' killer whales travel in pods of up to fifty and usually stay in the same place along the coast. They are quite predictable in their movements and their habits. 'Transient' pods are much less predictable and contain only about seven whales. They are rarely seen in the same place and tend to travel much farther afield than their 'resident' cousins.

• The Roman historian Pliny described the killer whale as 'a great mass of flesh armed with cruel teeth'.

4

A Family of Killers

With the exception of man, orcas are probably one of the most sociable and friendly creatures in the entire animal kingdom. They love the company of other orcas, and when they swim as a pod in the sea they like to keep as close to each other as possible. If a sole orca goes off on its own, it's never long before it returns to its pod. They need the presence and security of other orcas around them – which is why it's extremely cruel to keep just one sole orca in a tank.

In many marine theme parks orcas share their pools with other dolphins, but there's really nothing in the world to beat the company of members of their own kind, and most of them suffer from homesickness to one degree or another. The majority of marine parks have now at last realized this, but there are still several others which display just one lone orca. It's no wonder that Willy longed to be released from his pen and join his waiting brothers and sisters out in the open sea.

When Willy returned to the ocean he would have rejoined his pod (or family group). This is the basic social group of killer whales, and can be thought of as a sort of big extended family. There can be as many as fifty orcas in a pod, who travel, hunt and play together.

In a pod of fifty orcas you'd typically find that about ten (20 per cent) would be males, another ten would be calves, while the remainder would be females or immature males.

Orca society is very matriarchal. That's perhaps because female whales, although smaller in size, live much longer than males, and so attain a position of respect and authority within the pod. In fact, so dependent on their mothers are male orcas that it's not uncommon for them to die shortly after their mother has died – or been killed.

The average pod is made up of several smaller groups, called sub-pods. These sub-pods may sometimes leave the main pod for a period of time, but they always return to the larger group.

The orcas in a sub-pod are very closely related to each other; they could typically comprise mothers, daughters, sisters and cousins. They always travel together and are never separated – unless a human hunter captures some of them, of course. In fact, a mother and her calves will hardly ever be separated by a distance greater than 200 metres.

The pods and sub-pods make up much larger communities of perhaps 200 individuals who sometimes – but not always – travel together. Like the other social groups, the community is very strictly structured, and is effectively a closed society. Orcas from one community will never have anything to do with another community. Two communities of orcas have been discovered near British Columbia, a northern and a southern community. Despite the fact that they lived relatively close to each other, never once was an orca from the northern community seen socializing with one from the southern community.

Everyone in the pod knows his or her place, and the bond of loyalty and co-operation among the pod is extremely strong indeed. No one in a whale pod is alone, because each whale selflessly looks after every other whale, for the general bene-fit of the pod.

Two orcas have been known to tend a third sick orca, 'sandwiching' the sick orca between them to help him to stay afloat. Mothers prod their sick young, encouraging them to swim, and orcas will care for their sick and dying with a devotion that is impressive.

In one instance an old, wounded orca found itself beached on the shore. Every single member of its pod followed it, deliber-ately stranding themselves on the shore, pre-sumably trying to urge the old whale to return to the sea.

Despite the attempts of some human on-lookers to return the whales to the water, the creatures refused to move. It was only when the wounded whale died three days later that the rest of the pod went back to

their watery habitat. Fortunately none of the surviving whales was injured, even though some were severely sunburnt. This selfless co-operation of the orcas is in stark contrast to the way man often behaves. Perhaps we have something to learn from the orca after all.

The orcas fear only one creature – man. But even when attacked by man they are reluctant to escape individually and thereby split up the pod. This overriding sense of pod loyalty makes them oblivious to their own individual safety, or even just plain common sense. So if one orca is trapped in a net, the rest of the pod will hang around, concerned for its well-being. Of course, this makes it even easier for whale-hunters to trap them.

It shouldn't be thought that killer whales are helpless in the face of man, however. There are several reported incidents when orcas have successfully fought back and wreaked their revenge on those men who have harmed them in one way or another.

In British Columbia in 1956 a particularly heartless Canadian lumberjack deliberately released a log into the water so it would strike a member of a pod. The creature was stunned, and the lumberjack and his friend sailed off on their raft. That night when they returned to camp the orcas were waiting for them and overturned their raft. It's also said that the orcas only attacked the lumberjack who had released the log, leaving his innocent colleague alone.

In the first century AD an orca was stranded in the harbour at Ostia in Italy. The Roman Emperor Claudius ordered his Praetorian Guard to attack the hapless mammal with darts and javelins, 'for a pleasing spectacle for the people of Rome'. The orca was eventually killed, but not before it had succeeded in overturning at least one of the soldiers' boats.

Even farther back in history the philosopher Aristotle (who lived over two thousand years ago) tells of an orca which was wounded and captured by a gang of men. The other members of the captured orca's pod swam to shore and stayed there until the men took pity on the whale and released it.

Orcas stick together throughout their lives. Indeed the only time they are separated is when one or more members of a pod are captured by man.

Orcas may be one of the most caring and co-operative groups of mammals to be found anywhere in the world but, as their most commonly known name implies, they can also be deadly hunters. It's precisely their ability to work together as a team for the good of the pod which makes them such efficient predators and killers. Like wolves on dry land, orcas will often hunt in packs. When confronted by a pack of hungry killer whales the hapless prey can only wish for a speedy death.

Killer whales will eat practically anything. Unlike the other toothed whales and dolphins who eat mainly fish, small crustaceans and krill, the main diet of the orcas consists of seals, penguins and fish. They have also been known to attack porpoises and other dolphins. Even much larger whales like the gigantic blue whales are not safe from attack by a pack of orcas. They're very wary of walruses, however: perhaps it's the sharp tusks of the grizzled old creatures that make the orcas think twice before attacking.

Their social structure and strict inter-pod co-operation make them ideal hunters. If at all possible, they try and attack younger prey, as that's much easier to catch. But up to forty killer whales will band together if their prey is numerous or especially large.

First of all, the group of killer whales will surround their victim, cutting off all means of escape. Very often they will also try to separate their prey from its young. And then they dart in for the attack. It's a terrifying sight. Imagine an underwater Tyrannosaurus Rex!

It's for no small reason that orcas are feared throughout the oceans. Yet it's a mystery why they *never* attack human divers, except under extreme provocation.

Smaller packs of orcas will also show the same level of co-operation in hunting down prey. At the South Pole, for instance, seals are a staple of their diet. Normally the orcas hunt amongst the ice floes of the Antarctic Ocean, and shatter the floes upon which the seals are resting. But it's also been known for two orcas to swim *under* an ice floe and tilt it up to such a steep degree that the seal falls off – and straight into the waiting jaws of a third orca. Such a level of co-operation and careful pre-planning indicate that the orcas are extremely intelligent marine animals.

Their strong social set-up is essential for orcas to function as successful hunters. If there is easy prey to be had, the message will swiftly be conveyed through the pod and then the entire community. Similarly, if there's danger about, the orcas will alert their fellow whales.

Some years ago, Norwegian whalers received an SOS from a deep-sea fishing fleet. A great number of orcas (probably an entire community) had descended upon the area in which the fleet was fishing. The orcas were eating the fish in the area, leaving hardly any for the fishermen.

The whalers sped to the spot in their gunboats to discover that the fishermen hadn't been exaggerating: the whole sea seemed full of feeding and hunting orcas. Finally they shot one single whale with a harpoon which was loaded with explosive, a particularly brutal means of killing, but one commonly used on the larger whales. The other orcas took the hint and swam away, leaving the fishing grounds to the trawler-men.

Later that same day the orcas returned and resumed their hunting of the fish. But this time there was one vital difference. The orcas fed off the fish which were swimming in the vicinity of the fishing boats; they stayed well clear of the whaling boats with their harpoon guns. Obviously the orcas who had witnessed the killing of their fellow whale by the harpoon had reported that news back to the rest of their community and warned them to stay away from the whalers' boats.

That in itself would be noteworthy, but it's even more remarkable when you realize that the only difference between the two sets of boats was the harpoon gun jutting over the side of the whaling vessels. That important and life-saving piece of information had been passed to the other members of the pod and, by staying away from the harpoons, the orcas were able to feast in perfect safety. The fishing fleet and the whalers couldn't do anything to stop them: the orcas had beaten man at his own game!

However, an orca's single-minded and ruthless pursuit of its prey can often prove to be its downfall. In Patagonia, on the South American coast, killer whales will often pursue seals into the shallow waters off the coast where they end up being beached, doomed to a slow and lingering death, unless other orcas come to their rescue.

In 1937 in British Columbia an orca was so determined to kill a yapping dog on the shore that it even leapt out of the water and flung itself on to the rock shelf where the dog was. It seems that orcas will stop at nothing – even danger to their own life – to make a kill.

Even though their methods of killing may seem brutal to us, it's simply part of the natural order of things for killer whales.

They kill only to eat, never for sport (unlike men), and the instinct for hunting is in their blood. It must therefore be unbearably frustrating – and unnatural – for orcas such as Willy to be confined in one relatively small marine park tank, where by and large they're fed only fish which have already been killed.

Killer Whale Fact File

• *'Moby Doll' was an orca which was harpooned in 1965 by a sculptor who needed it as a model for a sculpture he was making. But he took pity on the wounded creature and nursed it back to health until he finally gained its confidence.*

• *In the official diving manual of the US Navy the subject of orcas is broached. The manual advises: 'The killer whale has a reputation of being a ruthless and ferocious beast ... If a killer whale is seen in the area the diver should get out of the water immediately!'*

• *It's impossible to calculate just how much*

killer whales eat in the wild, but in captivity they can eat up to about eighty kilograms of food a day, usually dead and frozen fish. It's probable that in the wild they eat much more. Some marine parks do feed them live fish, and even salmon. Just like Willy, orcas love salmon as much as some humans love chocolate.

• An orca which was caught in the Bering Sea had thirty-two full-grown seals in its stomach when it was slit open. Another orca was found with three pregnant porpoises inside it, while a seven metre killer whale contained thirteen porpoises and fourteen seals!

5

Whistling in the Dark

Off the coast of New England on a warm
summer's night you might hear some
strangely haunting music gently floating out
from the sea. There in the distance you will
see a school of humpback whales singing
softly to each other. It's one of the most
magical experiences in the world, and people
have often wondered what the meaning of
the whales' song is. It's perfectly clear that
they're communicating with each other, but
mankind still hasn't cracked the code of
whale and dolphin song.

Killer whales, like their larger cousins,
the humpback whales, obviously need to
communicate with each other underwater.

They speak to each other in a series of whistles and calls, some of which are audible to man. We can't say what these different sounds mean (although scientists are making some attempts to understand them), but it would seem reasonable to assume that at least some of the sounds are used as a sort of vocal mark of identification between whales belonging to the same pod.

Of course, many animals communicate with each other. Birds alter their song to pass on basic messages to other members of their flock, and the 'dances' of bees and wasps indicate to the rest of the hive the location of nectar, for example.

What is remarkable about the 'language' of the orcas, however, is that they speak in distinct and recognizable 'dialects'. Just as someone from Yorkshire would speak differently to a person from Somerset, so individual pods of whales speak with individual sounds and 'dialects'.

Each individual pod of orcas uses a series of about twelve sounds which is quite distinct from the series of sounds made by any other pod. In fact, just by listening to the dialect of two individual whales, a scientist can tell instantly whether the two orcas are members of the same pod or not. It should, in theory at least, be possible to say whether an orca comes from Patagonia or from Canada, as their sounds would be so different and local to the orcas' individual habitats. You could even say that one whale speaks with a South American accent and the other with a Canadian accent!

Two different pods may, however, have a few sounds in common; the more sounds they share, then the more closely related to each other they are. It's estimated that these distinct dialects are passed on in the pod from generation to generation and may be many thousands of years old. So even before *Homo sapiens* – man – had learnt to speak, the killer whales were already communicating very effectively with each other.

The calls which make up these distinct dialects are often faster, of a higher pitch and more frequent when the orcas are excited, perhaps by the approach of some potential prey.

The larger whales and dolphins also communicate with each other, of course, but certainly not in any discernible dialect. In fact, such dialects are extremely rare in the animal world, and are found in only a few animals – including man.

Perhaps we aren't the only creatures on the face of the globe who have developed a language. It's doubtful that the orcas' language is as structured as ours is, but there's no doubt that it's highly sophisticated.

Orcas produce other sounds as well as the calls and whistles which make up their individual dialects. Like all the toothed whales, they also produce a series of clicks and rattles, and it is these clicks and rattles which enable them to make their way through the murky waters of the ocean and track down their prey.

Human beings have five senses – touch, smell, taste, sight and hearing. Whales have only four: they cannot smell. However, their taste-buds are well developed and they can taste differences in the purity and chemical content of the surrounding water, so their sense of taste can be said to do the job of the sense of smell.

Their most important sense, however, is that of hearing: they can distinguish sounds up to a frequency of 280,000 hertz (man can reach only 20,000 hertz!). While they do rely on their sight in the clearer waters of the ocean, in turbulent waters, or in the depths where it is more difficult for sunlight to penetrate, the orcas rely on a system known as echolocation to move around and hunt.

In the dark, murky waters the killer whale emits a series of clicks which it is able to focus into a narrow beam of sound. It sweeps the beam of sound from side to side in front of it, and reads the echoes which come back from it. In this way the orca receives a 'sound picture' of its immediate environment and can 'see' any objects which are in its path – and also any potential prey.

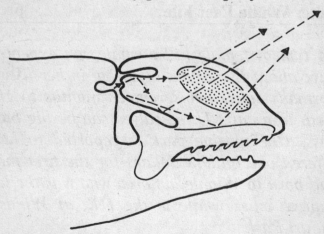

This is, of course, similar to the way bats navigate and see in the dark. Mankind has

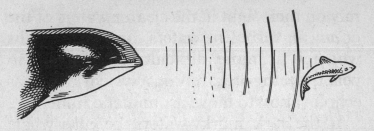

also developed a similar electronic system –
called sonar – which is used to locate
objects and vessels in the oceans' depths.
Sonar was first invented and put to use
in the middle of the twentieth century; Willy
and his fellow killer whales only beat mankind by approximately fifty million years!

Killer Whale Fact File

• *A typical female killer whale can expect to
have about five or six offspring in her life. It
normally takes her seventeen months to give
birth to a calf. These calves can be big busi-
ness. One marine park is reported to have
offered one million dollars for the first male
calf born to Winnie, an orca who was the last
captive killer whale in the UK, at Windsor
Safari Park.*

• *The first orca was put on display to the*

general public in 1964 on Vancouver Island, Canada. In 1862 London Zoo made its first attempt to keep a dolphin in captivity. It died the following day.

•The first orca born in captivity was a male calf born to whales Orky and Corky in 1977 in California. Unfortunately it died just two weeks later. Orky went on to have several more calves before he died in 1988, and some of his descendants are still alive today.

•Like elephants, killer whales are very receptive to training by humans, and they weigh almost the same as elephants. However, the size of a killer whale's brain is seven times the size of an elephant's brain!

6

Captured! Eight Shows a Day

The larger types of whale, such as the blue whale or the sperm whale, are a valuable source of many materials. They have been hunted and killed over the centuries for their blubber (which is, in turn, a valuable source of oil); their meat; their whalebone (which was used to make stiff, old-fashioned corsets); and spermaceti and ambergris, waxy substance which can be used to make cosmetics and perfumes. Thankfully, such ruthless exploitation of these endangered species has been more or less halted by international agreements.

Orcas, like Willy, were also exploited in a similar manner until 1982 when the International Whaling Commission insisted that there should be no more killing of the orcas until at least more was known about the impact such deaths had on the local orca population.

Orcas are still killed in small numbers as food for local people, and in Norway to stop them preying on herring, which is a valuable food crop for that country. It may still not be an ideal situation, but at least such limited hunting is a vast improvement on the thirty years before 1980 when about 6000 orcas were killed for their meat and oil.

Nowadays, the orcas are hunted and exploited in other far more subtle ways. This exploitation has become a major million pound industry, run not by grizzled old whalers but by slick and professional international businessmen.

Unlike the larger whales, the killer whales and dolphins aren't killed – although some do die accidentally – but are captured and brought to zoos, aquariums and marine theme parks all over the globe. There, they are exhibited to tourists and sightseers and taught to perform a series of tricks, solely for the entertainment of human spectators.

As many as eight times a day in some places, orcas are called upon to jump through hoops, perform somersaults in the water, even play aquatic football with each other or their human trainers, who will sometimes also ride around the pool on an orca's back.

Such tricks and stunts are certainly fun and exciting to watch, but people have recently been raising questions about the ethics of treating the orca just like any other performing animal. Just as more and more people are wondering whether it's right that circuses have performing animals, so they're also questioning man's right to treat orcas as yet one more tourist attraction. They're concerned about the welfare of the animals, the conditions they are kept in, and the ways in which they are captured in the first place.

Before an orca can be called on to perform its tricks it must first of all be captured and trained. Although orcas are found all over the world, most of the orcas captured today are from the north-western coast of the United States or from the waters of Iceland.

Over the centuries mankind has used many effective methods to capture orcas. The marine mammals have been caught in huge nets or, even more cruelly, harpooned: the Japanese were still harpooning orcas as recently as the 1980s.

One of the most common ways of catching orcas is by driving an entire group of them into a small bay or harbour. Once the mammals are inside, a net is dragged across the opening of the bay, thereby trapping them and enabling the whalers to pick them off one by one. If the orcas show reluctance to enter the bay, whale-hunters have even been known to detonate bombs to drive the panicking orcas into the ambush.

Another method is to throw tons of herring directly in the path of an oncoming pod. As the orcas greedily gulp down the food, a strong net is drawn around them, a net from which few, if any, can escape.

As we've already seen, orcas are so loyal to each other that even if only one creature is captured, the other members of the pod may stay around in the futile and certainly naïve hope of somehow helping their trapped fellow orca.

Of course, this only goes to make the hunter's task even easier. Even if just one orca is captured, the consequences for the rest of the pod can be potentially disastrous.

As most of these orcas are destined to end up in marine theme parks, performing tricks for the amusement of the general public, whale-hunters prefer to capture calves, rather than fully grown specimens. Apart from the relative ease of transporting the smaller calves, it's also been shown that younger orcas learn much more rapidly and easily than their older brothers and sisters. They also respond to captivity better than their elders, who have been used to the massive freedom of the oceans all their lives.

Assuming that the orca has not been killed while trying to avoid captivity, or being lifted aboard the whaler, there are still more dangers facing it before it reaches its destination.

Shipping (and flying) whales all over the world can be life-threatening. There has been one horrible report of an orca which was hoisted out of the water by a mechanical crane. As it was being raised out of the water, its weight was so great that it began to swing back and forwards, and was bashed constantly against the hard metal hull of the ship until it died a painful and agonizing death.

Even though there are now some internationally recognized guidelines on transporting orcas, the stress of being taken out of the ocean can be just too much for the orca, and many do still die *en route* to their new home.

How would you like it, for instance, if you were wrenched from the heart of your family and taken thousands of miles away from your home to a strange new environment? It can sometimes seem a wonder that captured orcas survive at all.

The first orca in captivity was displayed on Vancouver Island, Canada, in 1964, and since then the number in captivity has steadily grown, mainly in the USA, although one can find captive orcas in places as far apart as Hong Kong, Spain, France and Japan. At the time of writing there are no orcas in captivity anywhere in the UK, but this hasn't always been the case: the last orca in Britain was shipped to Florida as recently as the early 1990s.

Keeping an orca in captivity causes many problems for the animal. Orcas are very social creatures, as we've seen, constantly needing the companionship of other orcas. To keep just one orca in captivity seems to be barbaric, but there are still many marine parks which do just that. It's hardly surprising that upon arriving at its new home the captured and confused orca will refuse to eat food for several days; like Willy, it's pining for its family.

The threat of starvation will finally force the orca to eat, even though much of its food will consist of frozen and dead fish (very often with vitamin supplements). Orcas are used to hunting for their food and this new diet is undeniably a stark contrast to the tasty and live meat that makes up its steady diet in the ocean. It's a bit like being used to having fresh smoked salmon every day, and then to be suddenly told that from now on you'll have to be satisfied with nothing but frozen fish fingers.

So dazed and disorientated can the orca become that very often it has great difficulty in swimming normally for its first few days in captivity. It's not unusual to see a newly captured orca swimming round and round in circles, frightened and unsure in its new surroundings.

The orca might also lose its sense of balance and stability and behave as though it were drunk, finding it difficult even to swim in a straight line. There's also the additional hazard of the orca choking on items of food thrown into its pen by well-meaning visitors, which is why you'll find signs warning you against feeding the animals in zoos and safari parks up and down the country.

Certain guidelines have been drawn up in different countries about the minimum size of the pool in which the orca lives. (The UK has the strictest guidelines in the world – which may perhaps explain why there are no longer orcas in captivity in Britain.)

For instance, in the USA the length of a pool, or tank, which contains two orcas must be at least fourteen and a half metres, which is roughly twice the length of a fully grown orca, and four metres deep (approximately equal to the deep end of your local swimming pool). There are, of course, many tanks which are much larger.

Obviously, this is still a cramped space: in human terms it's roughly equivalent to a six-feet-tall man living in one single room about twelve feet square for the rest of his life.

It seems even more restricting when you consider that once the orca had the whole of the ocean as its 'tank'. Orcas have been known to cover 100 miles in a single day in the open sea, although a distance of thirty miles a day is probably more commonplace.

That's a little difficult to do when you're in a pool only about fourteen metres long! No matter how well looked after the orca is, it would be foolish and irresponsible to think that it doesn't get frustrated in such a comparatively confined space.

As we've already observed, orcas are very intelligent and sophisticated creatures and get on extremely well with human beings. They're also very receptive to training and learn very quickly to do the tricks required of them.

There are few things more thrilling than to see a fully grown orca leap several feet up out of the water to take a fish held out for it by its trainer, or to watch as it takes its trainer on a high-speed circuit of the pool. It's no wonder that the orcas and the dolphins are the biggest draw at all marine theme parks.

The orcas are trained by a system of performance and reward. Every time they do a trick correctly they're presented with a reward of food – usually dead fish such as herring – and gradually a rapport will grow between the orca and its trainer.

People have suggested that some trainers may refuse to give the orca food for days on end if it doesn't perform well and thereby starve it into giving a good show; but it seems that there's very little evidence of this type of starvation, and all orcas are fed every single day whether they carry out their tricks or not.

An orca may be required to perform up to eight times a day and so it's essential that the trainer builds up a level of trust with the orca. It's a shame that many trainers are young and inexperienced at handling orcas and dolphins; very little training is given, and some may have only relatively short-term contracts with the marine park they work for, so it's more difficult to create a bond of understanding between orca and human.

Having said that, there are indeed some very good and highly skilled trainers throughout the world who care deeply for their charges. Like Jesse and Willy, some of them have a strong empathy with each other, at times even seeming to know exactly what the other is thinking.

In recent years many of these trainers have spoken out about the conditions in which orcas are kept, and former trainers are some of the most ardent supporters of freeing all captive orcas and allowing them to roam their natural habitat once again.

However, no matter how much trust is built up between orcas and humans, things do sometimes go disastrously wrong. Orcas have a reputation for not attacking human beings unless provoked (which is strange because anything else in the ocean seems to be fair game for a hungry orca!), but there have nevertheless been very rare occasions when orcas have been known to attack their trainers, in some cases mauling them or even killing them.

There is no conclusive evidence as to what causes these rare attacks, although some could be put down to playfulness on the part of the whale. For instance, he might take his trainer in his mouth and swim down to the bottom of the pool with him or her, not realizing that human beings are not as adept as whales at surviving underwater for any length of time. For the orca it's nothing more than a game, albeit a deadly one.

Perhaps even more alarming are the reports of captive orcas attacking each other. Admittedly, these instances are very rare indeed, but even so this is something which never, ever, happens in the wild.

Some people who are firmly against the idea of orcas being held in captivity at all suggest that incidents like these may have something to do with the orca being taken out of its natural environment. After all, they were not born to be what they regard as little more than very highly skilled circus performers; they're wild animals, separated often by thousands of miles from their close-knit family and home, and natural hunters who no longer have the chance to do what they were born to do.

Other people have also suggested that orcas simply do not like to be ridden by humans, although the many stories down the centuries of humans safely riding dolphins would seem to disprove that theory.

Whatever the real reasons, some of these attacks on humans have been so horrific that several marine parks have now forbidden their trainers to ride the orcas, and prefer them to stay out of the water as much as possible. After all, it's better to be safe than sorry. Still, a man sharing a pool with a mature orca is probably many times safer than that same man in a lion's cage!

In the film *Free Willy* Jesse and Willy got on very well indeed, and no doubt enjoyed each other's company immensely. But in the end Jesse decided that Willy would be happiest in the sea again, reunited with his family.

So are orcas really happy performing their tricks and stunts for the members of the public who have paid to come and see their antics at a marine park? It's very hard to say. No doubt as very social creatures the orcas enjoy the company of their trainers, and the steady supply of food is particularly welcome, especially as there is never any guarantee of it in the wild.

It's certainly in the interests of the owners of marine parks to ensure that their orcas remain as healthy and happy as possible. They represent a considerable financial investment on their part: an orca can cost hundreds of thousands of pounds just to buy, and will attract large numbers of fee-paying tourists every year.

A female orca in the wild lives to an average age of fifty (although she can live to be up to ninety years old). A male has an average age of twenty-nine years, with a maximum lifespan of sixty years. As the first orca went on display only about thirty years ago it's too early to say whether female orcas in captivity can live to the age they do in the wild. However, it is a fact that of all the male orcas in captivity, only one has lived to the average age for males in the ocean.

In the wild orcas have a very varied diet –

everything from fish to seals, penguins and, as we've seen, other whales and dolphins. In captivity, however, they have a very limited diet of only a few kinds of fish. Most of the fish they are fed are dead; for an orca used to feeding on live meat, this may take some time to get used to.

The water quality in marine park pools is on the whole good. This is because the water in them is mixed with chlorine, the same chemical used in swimming pools to keep the water as clean and germ-free as possible. This has its own problems, however, and there are several instances of orcas developing severe eye infections and irritations from the amount of chlorine in the water. Possibly the best sort of pool for an orca is one into which sea-water is pumped. After all, this is the sort of water the orca is used to swimming in in the wild.

Captive orcas have been proved to have a lower activity rate than they do in the wild, and are often to be found listlessly swimming about their pool. This is hardly surprising as there's very little for them to do. Perhaps that's why – to spectators at least – they seem to enjoy leaping through hoops or playing games with each other. At least it's something for them to do.

Even though there is very little to occupy them, they still suffer from stress; some even develop ulcers! It's been suggested that one cause of stress, apart from their restricted freedom, might be the material from which the walls of their tanks are made.

We've learnt that, to a large extent, an orca 'sees' with its ears, sending off sound signals which rebound off any object directly in its path. It's possible (although it hasn't been proved) that the walls of their tanks reflect the sounds in such a way that the orca becomes totally disorientated. To understand this better, try and remember how you feel when you're confronted with multiple and distorted images of yourself in the hall of mirrors at a funfair. The orca would feel just as confused and dislocated in a tank.

The most important factor in an orca's life is its pod, and the members of that pod. The whales in a pod play together, socialize and hunt with each other, and the bond between members of each pod is probably the strongest relationship of its kind to be found amongst any creatures in the entire animal world – and that includes man. But nowhere in the world is there a complete pod of orcas in captivity, and many of the whales have no hope of ever being reunited with their families. It must be a terrible traumatic shock for many of them.

The great dorsal fin on the back of a male orca is an impressive sight as it slices through the waters of the world's oceans. This sword-like fin is often as much as six feet high, and is firm and erect. But if you expect to see this impressive fin on an orca in captivity you're due for a big disappointment. Once in captivity the dorsal fin of the male orca begins to droop until it eventually collapses.

We don't know why this is so. It could be a sign of ill health. On the other hand, as the orca uses its dorsal fin as a stabilizer in the ocean, it might just mean that in the relatively calm environment of its tank it simply doesn't need it any more. Whatever the reason, there's no escaping the fact that a collapsed dorsal fin is simply not natural.

So are orcas happy in captivity? The short answer is that, until we devise some means of communicating with orcas, we may never really know for sure.

Killer Whale Fact File

• *Whale trainers make a great show of putting their head into an orca's open mouth. To date no orca has ever clamped his jaws shut over the head of his trainer!*

• *At the time of writing orcas are on public display at seventeen marine parks throughout the world. The majority of these are in North America, but orcas can also be seen in the south of France and in Barcelona, Spain. Six of these marine parks hold only a single orca. At the others the orca is accompanied by up to four other members of the same species.*

• *Mankind has hunted whales since the beginning of time. The first record of whaling we have has been found on the walls of prehistoric caves in Norway, indicating that the people who live by the cold waters of the north were probably the very first whalers. The modern tradition of whaling seems to have begun about the*

70

year 1100 among the Basque people of southern France and northern Spain. On sighting a whale they would put out to sea and, with blazing torches, would drive the confused and frightened whale on to the shore. It's even been suggested that the word 'harpoon' might come from the old Basque word arpoi which means 'to catch alive'.

• One of the greatest novels in world literature is Moby-Dick by the American writer Herman Melville. It tells the story of Captain Ahab's hunt for a huge sperm whale in the nineteenth century. Even though it's a work of fiction, it has never been equalled for its information on the larger whales and whale-hunting.

7

Allies of the Deep

Beyond any doubt whatsoever, whales and dolphins, and the orca in particular, are amongst the most intelligent creatures on the surface of this planet. We know this from their complex family set-ups, their loyalty to each other and their use of language. It takes brains to plan and organize a hunting party to ambush a sea-lion, or tip a resting seal from its ice floe into the waiting mouth of another killer whale. Only an exceptionally intelligent species would be able to recognize danger in the shape of harpoon boats, for instance, and be able to convey exact details of that threat to the other members of its pod.

Similarly, their inquisitiveness suggests an extremely lively and inquiring mind. Dolphins especially have been known to swim up to ships just to see what's going on, and will often accompany ships for many miles. This insatiable curiosity can often be their downfall, as they are then easily picked off and caught.

Their ready receptiveness to training and their ability to perform tricks alone proves that they have some minimal intelligence. Just how clever they are we may never know, but scientists are confident that on the intelligence scale orcas lie, at the very least, somewhere between a dog and a chimpanzee. The reality may be even more surprising, but until we devise some reliable means of communicating with these creatures we can never be sure.

All orcas seem to have a peculiar affinity with man, no matter how brutally human beings may have treated the species in the past – and are indeed still doing today. The noises of ships' engines seem to attract them (remember, their sense of hearing is very acute indeed), and it's not unusual at all for

a pod of orcas to accompany steamboats
fishing for cod or herring off the coast of
Iceland. They know that where the steamboats go there will also be a ready supply of
food for them.

Willy saved Jesse's life when the young
boy fell into the killer whale's tank, and
instances of orcas and dolphins rescuing
people from death at sea are not at all
uncommon.

The case of Yvonne Bliss is a famous

case in point. In March 1960, fifty-year-old Mrs Bliss fell off the deck of a ship which was cruising down the Bahama Channel in the West Indies. Although a swimmer, she soon found herself in difficulty in the cold and choppy waters and, as she struggled to stay afloat, the boat was speeding ever farther away from her.

The currents in that part of the sea are strong, and Mrs Bliss was finding it difficult to breathe, as wave after wave slapped against her side, stinging her eyes, and filling her mouth with salt-water. Then the battering from the waves suddenly stopped. A dolphin – and its behaviour strongly recalls that of its orca cousins – had swum up beside Mrs Bliss, placing its long body in between her and the current, thereby stopping her from taking in water.

As Mrs Bliss recovered her breath, and her level-headedness, the dolphin slowly steered her out of the turbulent currents and towards shallower and calmer waters, before heading off back to the ocean. The dolphin had, in short, selflessly saved the life of a complete stranger who – what's more – belonged to a totally different species!

There are numerous other similar instances which seem to prove that the legends we talked about earlier in Chapter 3 do have a grain of truth in them. Killer whales, like Willy, seem readily to accept man as a sort of honorary orca, and accord to him the same sort of support and help they would give to another member of their pod.

We have absolutely no idea why this should be so – after all the bad treatment we've inflicted on whales, orcas and dolphins down through the centuries, one would have thought that man would be the last animal an orca would want to help out – just as we don't know for certain why a hungry orca will attack a porpoise or another whale but never a human being.

Perhaps the whales, out there in the ocean and in their tanks in marine parks, know something that we don't?

As we come to understand orcas more, and recognize their intelligence and their co-operative spirit, so scientists are examining the various ways in which orcas can help mankind.

In some parts of Greece fishermen still

call on the help of dolphins to drive fish into their waiting nets, a practice which has been carried out for thousands of years.

Many scientists believe that orcas could help mankind enormously in his conquest of the sea. We commonly think of space as the 'final frontier' which mankind has to cross but, in reality, the ocean depths are just as foreboding and mysterious as the cold wastes of space. And mankind might find himself having to return to his exploration of the sea sooner than he imagines.

The earth is approaching a population crisis. At the beginning of this century the world's population was a little over 1½ billion. In 1987 it was 5 billion, and by the turn of the century it's been estimated that there will be over 6 billion people living on the surface of this planet. That figure is just under 10 per cent of all the people who have ever been born on earth. Clearly, if the increase in population continues at this rate, sooner or later there is not going to be enough food to go round.

But there is one place on earth where the supplies of food are virtually limitless, and it's also an area whose potential has hardly been exploited at all. The whales and the dolphins have known about it ever since their ancestors left the land over fifty million years ago: the sea.

It's not at all inconceivable – indeed, it's probable – that mankind may one day start farming the sea. And just as a farmer on land needs the help of beasts of burden, so undersea farmers could find the help of marine animals invaluable. Dolphins and orcas would then be set to become man's underwater allies, with the six-ton orca being able to pull along behind him enormous weights.

Orcas might also take the role of a sort of St Bernard of the sea, rescuing divers who encounter difficulties, as this is something which seems to come naturally to orcas. Using echolocation, orcas could even guide to safety divers who have become lost and disorientated in the dark and murky waters at the bottom of the ocean. Few creatures would dare attack a mature orca, so orcas could even become mankind's underwater 'guard dogs', and they would also be able to swim into places inaccessible to man.

All these roles for orcas have been suggested and considered by various governmental and voluntary bodies. More sinister uses for orcas have also been suggested, such as employing them to attack the divers of an enemy power; although, given their natural aversion to attacking men, it's not certain whether orcas could be trained to do this.

However, something very similar has already happened. During the 1960s the US Government used orcas to retrieve torpedoes – a potentially hazardous operation – and to carry and attach limpet bombs to enemy ships. There have even been accusations that dolphins were trained to carry and plant undersea mines during the Gulf War of the early 1990s.

It would be a great pity, however, if man used the noble and highly intelligent killer whales as just another weapon with which to settle his petty quarrels and squabbles.

Killer Whale Fact File

•*In the 1950s the US Navy destroyed hundreds of orcas with mines and depth charges, claiming that the killers were posing a hazard to their ships.*

• *Up to the 1930s Siberians were feeding their dogs with whale blubber which had been lying far above sea-level, and which had been preserved in the sub-zero temperature for 100,000 years!*

• *Although not a numerous species, killer whales are not considered an endangered species. This is in stark contrast to their blue whale cousins. Since the beginning of this century hundreds of thousands of blue whales have been slaughtered in the open sea. There is now a very real danger that this species could become extinct.*

Epilogue

Killers which have never harmed a single human being without provocation; 'fish' that are really mammals; dreaded and supposedly bloodthirsty hunters of the sea, who are much more likely to guide you safely in to shore if you're experiencing difficulties in the water; and 'dumb' performing animals which many believe are the most intelligent creatures on this planet after ourselves. Willy and his fellow orcas are surely one of the most intriguing animals known to man.

For centuries we've exploited the orca and its dolphin and whale cousins to such an extent that many species of the larger whales are now in danger of becoming as extinct as the dinosaurs. Most of the brutal whale-hunting of the past has now been outlawed by international treaties, but many people fear the whales will never really recover. The blue whale, the largest creature that has ever existed, now numbers only two thousand in the entire world, when once it swam the ocean currents in hundreds of thousands.

Thankfully, Willy and his fellow orcas are in no danger of becoming classified as an endangered species, at least not for the time being. Yet we still take individual orcas from their closely knit pods and keep them in captivity, often without any other orcas for company, and frequently with a less than adequate concern for their well-being.

There are two main arguments for keeping animals in captivity, whether it be an orca in a tank, or a lion in a cage at the zoo. One is that the animal is so threatened in the wild by natural predators that it's in danger of becoming extinct. By keeping it in a controlled and protected environment, so the supporters of this argument say, we can ensure that at least one more species doesn't die out. However, orcas as yet are in no danger of becoming extinct and, as one of the most efficient and ruthless killers in the ocean, they need fear no other animal apart from man.

The other argument maintains that animals in captivity can be a valuable educational tool, giving people the opportunity of observing them at close quarters, something which they normally wouldn't have the chance of doing.

Apart from the fact that the educational value of orcas in captivity is limited – after all, orcas in a tank behave in a totally different way to how they behave in the ocean – the fact remains that orcas remain in captivity largely, and often exclusively, for the entertainment value they hold for the tourists who come to visit them every year.

Going through the routine of the same tricks each and every day, the orcas are little more than the aquatic version of the clowns you might see in the circus ring. The only difference is that at least the clowns at the circus choose to perform; the orcas have no say whatsoever in the way they are made to live the rest of their lives.

Undoubtedly, visitors to marine parks will gain a thrill from seeing a live orca, and perhaps even learn something from the experience. But how much more educational it would be to see orcas in the wild. With today's technology we don't have to travel to the North Sea or the coast of British Columbia to see killer whales in their natural habitats.

Modern-day video and satellite technology can give us a much more vivid, exciting and genuine picture of orcas in the wild than any sole 'showbiz' orca in the false environment of its marine park tank. Some marine parks – including one run by the family of the famous French oceanographer Jacques Cousteau – have realized this, and their audio-visual displays are just as thrilling as seeing an orca in the wild.

There are still lots of people who don't accept this argument. They say that a killer whale is just another animal – awe-inspiring, certainly; beautiful even; but Willy and his friends are just another example of dumb beasts. Aren't they?

Well, not quite. We know that killer whales and dolphins have a very sophisticated social set-up, and that they can communicate with each other in a way in which few other animals are able to. Like Willy with his close relationship with Jesse, they also have a curious affinity with human beings, and have even been known to save people's lives.

They can't understand English (or Latin-American Spanish, or Japanese for that matter, even though they're frequently seen off the coasts of Argentina and Japan), but they are able to follow simple instructions from their trainers in marine parks and, in controlled scientific experiments, understand quite complex sentences in sign-language.

Yet not one human being has managed to understand the language of the orcas. Orcas have evolved in an environment which couldn't be more different to our own environment; their intelligence, their language and perhaps even a culture might also have developed in a completely different way to ours, in a way which we can't even comprehend. People may think that orcas are just dumb beasts because we can't communicate with them; perhaps orcas are thinking exactly the same thing about us!

Intelligent, caring and beautiful, orcas, whales and dolphins have conquered almost three-quarters of our planet, leaving only the final quarter – dry land – to man. In evolutionary terms they are our very long-distant cousins, and perhaps we should regard them not as cute circus performers but as our fellow mammals and our equals.

Perhaps the orcas can then safely swim the oceans of the world, secure in the knowledge that they will never again be captured and taken away from their pods, and that orcas like Willy and his family will always be free.

The Whale and Dolphin Conservation Society is a voluntary organization which studies the welfare of whales and dolphins both in captivity and in the wild. They publish a wide range of material about whales and dolphins. They also run a scheme whereby you can adopt a whale or dolphin in the wild.

If you want to help the Society, or are interested in the work they do, please write, enclosing a stamped addressed envelope, to them at:

The Whale and Dolphin Conservation Society
19A James Street West
Bath
Avon BA1 2BT

Also in Fantail

ROBIN HOOD PRINCE OF THIEVES

by Simon Green

The legend lives on. Like a flaming arrow, Robin of Locksley
emerges from the shadows of Sherwood Forest to blaze a path
for the poor and downtrodden. With a mighty band of fighting
men by his side – Friar Tuck, Will Scarlet, the noble Saracen
called Azeem, and others – Robin wages a magnificent war
against the vicious Sheriff of Nottingham . . . and an equally
passionate campaign for the heart of the beautiful Maid Marian.
Wielding his bow and arrow with deadly accuracy, Robin of
Locksley transforms himself into a new kind of hero

THE ADDAMS FAMILY

by Elizabeth Faucher

You haven't lived unless you've met the Addams Family! There's
Morticia, the loving, caring mother, Gomez, the devoted but
manic father, and their children Pugsley and Wednesday. Pugsley
collects road signs, and Wednesday's favourite toy is a headless
doll. Then, of course, there's Thing, the Addams Family's pet
hand, who is always willing to lend one, when two just aren't
enough. With the return of Uncle Fester, the long-lost brother of
Gomez, after twenty-five years' absence, the family is complete
once again. However, he may look like Uncle Fester, he may even
sound like him, but can he really be the missing uncle?

FANTAIL